The Groundbreaking Story of the First-Ever Pig Kidney Transplant in saving man's life

How One Man's Brave Journey Offers Hope for Thousands in Need of Life-Saving Organ Transplants

Gardner Tenny

Copyright © 2024 By Gardner Tenny

All rights reserved. No part of this publication may be reproduced, distributed, or transmitted in any form or by any means, including photocopying, recording, or other electronic or mechanical methods, without the prior written permission of the publisher, except in the case of brief quotations embodied in critical reviews and certain other noncommercial uses permitted by copyright law.

TABLE OF CONTENTS

Introduction   4
Richard Slayman's Transplant Journey   15
Why Pig Organs Are Considered for Transplants 23
   Exploring the Potential and Challenges   24
Overcoming Rejection   32
   Advances in Transplant Techniques   32
   Brighter Future for Organ Transplantation   41
Massachusetts General Hospital   43
   Pioneering Transplant History   43
Genetic Modifications of Pig Donors   53
   Revolutionizing Xenotransplantation for Human Compatibility   53
Understanding Richard Slayman's Medical Background   62
   A Journey of Hope and Resilience   70
Prognosis After Transplant   72
   Navigating the Journey to Recovery and Beyond   72
   A Journey of Hope and Healing   81
Comparing Success Rates with Previous Attempts   82
   Advancements in Transplantation   83
Future Possibilities   92
   Transforming the Landscape for Dialysis Patients   92
Cost and Approval Considerations in Advanced

Medical Therapies	**102**
Looking Ahead in Transplant Medicine	**113**
Acknowledgment	**121**

Introduction

Organ transplantation has long been a critical medical procedure, offering hope to countless individuals suffering from organ failure. However, the demand for donor organs far exceeds the available supply, leading to lengthy waiting lists and, tragically, the loss of many lives. In recent years, researchers have been exploring innovative solutions to address this pressing issue, and one breakthrough stands out: the use of genetically modified pig organs for transplantation.

1. The Need for Organ Transplants

Organ transplantation is a life-saving treatment for individuals with end-stage organ failure. Whether due to disease, injury, or congenital conditions, the failure of vital organs such as the heart, kidneys, liver, and lungs can have devastating consequences. For many patients, a transplant represents their only chance at survival and a return to a normal, healthy life.

Despite the increasing success of transplantation procedures, there remains a critical shortage of donor organs. This shortage is exacerbated by factors such as strict eligibility criteria, limited availability of suitable donors, and logistical challenges in organ procurement and allocation. As a result, thousands of patients languish on waiting lists for months or even years, their health deteriorating as they await a compatible donor organ.

2. Challenges in Organ Transplantation

The scarcity of donor organs is just one of the challenges facing the field of organ transplantation. Another major hurdle is the risk of organ rejection, wherein the recipient's immune system recognizes the transplanted organ as foreign and mounts an immune response against it. This can lead to graft failure and necessitate additional medical interventions, including the administration of immunosuppressive drugs to suppress the immune system and prevent rejection.

Moreover, the limited lifespan of donor organs poses a significant obstacle to long-term transplant success. Even in cases where the transplant is initially successful, organ function may decline over time due to factors such as chronic rejection, infection, or recurrence of the underlying disease. As a result, many transplant recipients require multiple transplants throughout their lifetime, further straining the already constrained organ supply.

3. The Promise of Genetically Modified Pig Organs

In light of these challenges, researchers have been exploring alternative sources of donor organs that could help alleviate the shortage and improve transplant outcomes. One particularly promising avenue of investigation involves the use of genetically modified pig organs for transplantation into humans.

Pigs are an attractive candidate for organ donation due to their physiological and anatomical similarities to humans, as well as their rapid breeding and high organ yield. However, early attempts at xenotransplantation, or the transplantation of organs between different species, were hampered by the rapid rejection of pig organs by the human immune system.

Recent advances in genetic engineering have paved the way for the development of genetically modified pigs with enhanced compatibility and reduced immunogenicity. By modifying the pig's genome to eliminate or modify certain genes involved in the immune response, researchers have been able to create pigs whose organs are less likely to be rejected when transplanted into humans.

4. Breakthrough Transplantation Procedures

The successful transplantation of genetically modified pig organs into human recipients represents a significant milestone in the field of xenotransplantation. These groundbreaking procedures offer hope to patients who have exhausted all other treatment options and are facing the prospect of irreversible organ failure.

One of the most notable examples of this achievement is the recent transplant of a genetically modified pig kidney into a human recipient at Massachusetts General

Hospital. This pioneering procedure, performed by a team of skilled surgeons and researchers, demonstrated the feasibility and safety of using pig organs for transplantation in humans.

The recipient of the pig kidney, Richard "Rick" Slayman, had been suffering from end-stage renal disease and was facing a bleak prognosis without a transplant. However, thanks to the innovative efforts of the medical team at Massachusetts General Hospital, he was given a new lease on life with the successful implantation of the pig kidney.

5. Implications for Organ Transplantation

The successful transplantation of genetically modified pig organs has far-reaching implications for the field of organ transplantation and for the millions of patients worldwide who are in need of life-saving treatment. By expanding the pool of available donor organs, these procedures have the potential to reduce waiting times, improve transplant outcomes, and save countless lives. Furthermore, the development of genetically modified pigs for organ donation could help address some of the ethical concerns surrounding organ transplantation. Unlike human donors, pigs can be bred specifically for organ donation, alleviating concerns about exploitation and the equitable distribution of donor organs.

However, despite the promise of xenotransplantation, numerous challenges remain to be addressed before pig organs can be routinely used in human transplantation. These include concerns about the risk of cross-species transmission of infectious diseases, the long-term safety and efficacy of pig organs in humans, and the ethical implications of genetically modifying animals for organ donation.

The successful transplantation of genetically modified pig organs represents a major breakthrough in the field of organ transplantation, offering hope to patients who are in desperate need of life-saving treatment. By harnessing the power of genetic engineering, researchers have overcome longstanding barriers to xenotransplantation and paved the way for a new era of transplantation medicine. While significant challenges remain to be addressed, the promise of genetically modified pig organs holds tremendous potential to transform the landscape of organ transplantation and improve the lives of millions of patients worldwide.

With continued research and innovation, we may one day see a future where organ shortages are a thing of the past, and every patient in need has access to a compatible donor organ.

Richard Slayman's Transplant Journey

Richard "Rick" Slayman's transplant journey is a testament to the resilience of the human spirit and the transformative power of medical innovation. His experience highlights the challenges faced by individuals with end-stage organ failure and underscores the importance of exploring novel approaches to organ transplantation, such as xenotransplantation using genetically modified pig organs.

1. Early Life and Health Challenges

Richard Slayman was born and raised in Weymouth, Massachusetts, where he led a relatively normal life until his health took a sudden turn for the worse. In his early thirties, Rick was diagnosed with type 2 diabetes, a chronic condition characterized by high blood sugar levels. Despite managing his diabetes with medication and lifestyle changes, Rick's health continued to deteriorate over time.

2. Diagnosis of End-Stage Renal Disease

As Rick's diabetes progressed, it began to take a toll on his kidneys, eventually leading to end-stage renal disease (ESRD). ESRD occurs when the kidneys fail to function adequately, resulting in the accumulation of waste products and fluid in the body. Without intervention, ESRD is fatal, requiring either dialysis or a kidney transplant to sustain life.

Rick's diagnosis of ESRD was a devastating blow, requiring him to undergo regular dialysis treatments to remove waste products from his blood. Dialysis, while life-saving, is a burdensome and time-consuming procedure that significantly impacts a patient's quality of life.

3. The First Kidney Transplant

In 2018, Rick received his first kidney transplant from a deceased donor, offering him a new lease on life after years of struggling with kidney failure. The transplant surgery was successful, and

Rick's new kidney initially functioned well, allowing him to resume many of his daily activities and enjoy a better quality of life.

However, despite the initial success of the transplant, Rick's health took a turn for the worse when his body began to reject the donor kidney. Despite medical interventions to suppress his immune system and prevent rejection, Rick's transplanted kidney ultimately failed, leaving him once again reliant on dialysis to stay alive.

4. Exploring New Treatment Options

With his first transplant failing and his health rapidly declining, Rick faced a daunting reality: without a new kidney, his prospects for survival were grim. Traditional organ transplantation from deceased donors offered limited hope due to the shortage of available organs and the high risk of rejection.

Faced with this dire situation, Rick and his medical team began to explore alternative treatment options, including the possibility of xenotransplantation using genetically modified pig organs. While still considered experimental, xenotransplantation offered Rick a chance at a new kidney and a renewed hope for the future.

5. The Decision to Proceed with Xenotransplantation

After careful consideration and consultation with his medical team, Rick made the courageous decision to undergo xenotransplantation with a genetically modified pig kidney. Despite the inherent risks and uncertainties associated with the procedure, Rick saw it as his best chance for survival and a way to pave the path for future patients in need of organ transplants.

6. The Transplant Surgery

On March 16, 2024, Richard Slayman underwent a historic transplant surgery at Massachusetts General Hospital, becoming the first person to receive a genetically modified pig kidney. The surgery, which lasted four hours, was performed by a team of skilled surgeons led by Dr. Tatsuo Kawai. The transplant procedure involved connecting the blood vessels and ureter of the pig kidney to Rick's own circulatory and urinary systems, allowing the organ to function within his body. Remarkably, the pig kidney immediately began producing urine, a promising sign of its viability and functionality.

7. Road to Recovery

In the days following the transplant surgery, Rick's condition steadily improved, and he began to show signs of recovery. Despite the inherent risks of organ rejection and other complications, Rick remained optimistic about his prognosis and focused on his journey towards healing and rehabilitation. With the support of his medical team, family, and friends, Rick embarked on a new chapter in his life, filled with hope and gratitude for the opportunity to receive a second chance at life through the groundbreaking transplantation of a genetically modified pig kidney.

8. A Beacon of Hope

Richard Slayman's transplant journey serves as a beacon of hope for individuals facing organ failure and underscores the transformative potential of medical innovation in saving and improving lives. His story highlights the importance of perseverance, courage, and collaboration in overcoming the challenges of organ transplantation and advancing the frontiers of medical science. As Rick continues his recovery and adjusts to life with his new kidney, his journey will undoubtedly inspire countless others and pave the way for future advancements in xenotransplantation and the use of genetically modified pig organs in human patients.

Through his resilience and determination, Rick has not only changed his own life but has also contributed to the collective pursuit of a world where organ failure is no longer a death sentence, but rather a treatable condition with the promise of new hope and possibilities.

Why Pig Organs Are Considered for Transplants

Exploring the Potential and Challenges

The shortage of human donor organs for transplantation has long been a critical challenge in the field of medicine, prompting researchers to explore alternative sources to meet the growing demand for life-saving transplants. Among the most promising alternatives are pig organs, which offer unique advantages and present novel opportunities for addressing the organ shortage crisis. In this comprehensive exploration, we delve into the reasons why pig organs are considered for transplants, examining the biological, logistical, and ethical factors driving their potential use in clinical transplantation.

1. Biological Compatibility and Similarity

Pigs, like humans, are mammals, sharing many physiological and anatomical similarities that make their organs compatible with human recipients. From the size and structure of organs to the functioning of vital systems such as the cardiovascular and immune systems, pigs closely resemble humans in many aspects, making their organs suitable candidates for transplantation. Additionally, pigs have a similar lifespan to humans, allowing for long-term studies and observations of organ function and compatibility in preclinical and clinical settings.

2. Availability and Abundance

Unlike human donor organs, which are scarce and limited in supply, pigs offer a readily available and abundant source of organs for transplantation. With billions of pigs raised and bred for food production worldwide each year, there is a surplus of organs that could potentially be utilized for transplantation, alleviating the critical shortage of donor organs and expanding access to life-saving treatments for patients in need. Moreover, the large size of pigs relative to humans makes their organs particularly suitable for transplantation into adult recipients, providing a potential solution for patients awaiting compatible donor organs.

3. Ethical Considerations and Public Acceptance

While the use of pig organs for transplantation raises ethical concerns and prompts debates about animal welfare and the ethics of xenotransplantation, public acceptance of this approach has grown in recent years. Advancements in genetic engineering and animal welfare practices have led to the development of genetically modified pigs specifically bred for transplantation, minimizing the ethical concerns associated with traditional farming practices. Additionally, public awareness campaigns, education initiatives, and discussions among stakeholders have helped foster greater understanding and acceptance of the potential benefits of pig organ transplantation in addressing the organ shortage crisis.

4. Immunological Compatibility and Genetic Modification

One of the major challenges in xenotransplantation, the transplantation of organs between different species, is the risk of rejection due to immunological incompatibility. Pigs possess certain antigens and proteins on the surface of their cells that can trigger an immune response in humans, leading to rejection of transplanted organs. However, recent advancements in genetic engineering have enabled scientists to modify pig genomes to reduce the expression of these antigens and proteins, making pig organs less immunogenic and more compatible with human recipients. By targeting specific genes involved in immune recognition and rejection, researchers aim to create genetically modified pigs with organs that are less likely to be rejected by the human immune system, thereby improving the success rates of pig organ transplantation.

5. Challenges and Future Directions

Despite the promising potential of pig organs for transplantation, several challenges remain to be addressed before this approach can be widely adopted in clinical practice. These challenges include the risk of infection from porcine viruses, the potential for immune rejection despite genetic modifications, and the ethical considerations surrounding the use of animals for transplantation. Additionally, regulatory approval, public acceptance, and healthcare infrastructure for xenotransplantation pose significant hurdles that must be overcome through collaborative efforts and continued research. Nevertheless, with ongoing advancements in genetic engineering, immunology, and transplantation medicine, pig organ transplantation holds immense promise as a viable solution to the organ shortage crisis, offering hope for millions of patients awaiting life-saving transplants.

Pig organs are considered for transplantation due to their biological compatibility, availability, ethical considerations, and potential for genetic modification. Despite the challenges and complexities associated with xenotransplantation, pig organs offer a promising solution to the critical shortage of human donor organs, providing hope for patients in need of life-saving transplants. As researchers continue to innovate and overcome the remaining hurdles, pig organ transplantation may soon become a reality, revolutionizing the field of transplantation and improving outcomes for patients around the world. Through interdisciplinary collaboration, ethical stewardship, and continued research, we can unlock the full potential of pig organs for transplantation and usher in a new era of hope, healing, and possibility in medicine.

Overcoming Rejection

Advances in Transplant Techniques

Organ transplantation has transformed from a novel medical procedure into a life-saving treatment for countless individuals suffering from end-stage organ failure. However, the success of transplantation is often hindered by the recipient's immune system, which can recognize the transplanted organ as foreign and mount an immune response, leading to rejection. Overcoming rejection is therefore paramount to the long-term success of organ transplantation. In recent years, significant advances have been made in transplant techniques aimed at minimizing rejection and improving transplant outcomes. This essay explores the various strategies and innovations developed to overcome rejection in organ transplantation.

1. Understanding Organ Rejection

Before delving into the advancements in transplant techniques, it is essential to understand the mechanisms of organ rejection. Organ rejection occurs when the recipient's immune system recognizes the transplanted organ as foreign and launches an immune response against it. This response can manifest as acute rejection, occurring within days to weeks after transplantation, or chronic rejection, developing over months to years. Both forms of rejection can lead to graft failure and necessitate additional medical interventions to salvage the transplant.

2. Immunosuppressive Therapy

One of the cornerstone strategies for overcoming rejection in organ transplantation is the use of immunosuppressive therapy. These medications work by suppressing the recipient's immune system, thereby reducing the risk of rejection and prolonging the survival of the transplanted organ. Common immunosuppressive agents include calcineurin inhibitors, such as cyclosporine and tacrolimus, as well as corticosteroids and antimetabolites.

While immunosuppressive therapy has revolutionized organ transplantation and significantly improved transplant outcomes, it is not without its drawbacks. Prolonged use of immunosuppressive medications can increase the risk of infection, malignancy, and other adverse effects, posing challenges for transplant recipients and their healthcare providers.

3. Targeted Immunosuppression

To mitigate the side effects of broad-spectrum immunosuppressive therapy, researchers have developed targeted immunosuppression strategies aimed at selectively suppressing the immune response against the transplanted organ while preserving overall immune function. These approaches include the use of monoclonal antibodies targeting specific immune cell populations, such as T cells and B cells, as well as novel immunomodulatory agents designed to modulate immune activity without causing global immunosuppression.

Targeted immunosuppression holds the promise of improving transplant outcomes and reducing the burden of immunosuppressive medications on transplant recipients. By selectively targeting the immune pathways involved in rejection, these therapies offer the potential to achieve better graft survival while minimizing the risk of infection and other complications associated with traditional immunosuppressive regimens.

4. Induction Therapy

Induction therapy involves the administration of potent immunosuppressive agents immediately before or during transplantation to prevent early rejection and promote graft acceptance. This approach is particularly beneficial for high-risk transplant recipients, such as those receiving organs from deceased donors or those with pre-existing sensitization to donor antigens.

Common induction agents include monoclonal antibodies, such as anti-thymocyte globulin and basiliximab, which target immune cells involved in the early stages of rejection. By providing robust immunosuppression during the critical peri-transplant period, induction therapy can improve short-term transplant outcomes and reduce the incidence of acute rejection episodes.

5. Tolerance Induction

In addition to preventing rejection, researchers are actively exploring strategies to induce immune tolerance towards transplanted organs, thereby obviating the need for lifelong immunosuppressive therapy. Tolerance induction aims to reprogram the recipient's immune system to recognize the transplanted organ as self and tolerate its presence without mounting an immune response.

Various approaches to tolerance induction are being investigated, including mixed chimerism, regulatory T cell therapy, and donor-specific tolerance protocols. These innovative strategies hold the potential to revolutionize organ transplantation by offering a cure for rejection and enabling transplant recipients to enjoy long-term graft survival without the need for immunosuppressive medications.

6. Biomarker Monitoring

Advances in biomarker monitoring have revolutionized the detection and management of rejection in organ transplantation. Biomarkers, such as circulating cell-free DNA, donor-specific antibodies, and gene expression profiles, provide valuable insights into the immune status of transplant recipients and the likelihood of rejection.

By monitoring biomarkers longitudinally, transplant clinicians can identify rejection early, allowing for timely intervention and management. Furthermore, biomarker-based surveillance enables personalized immunosuppressive therapy, optimizing transplant outcomes while minimizing the risk of over-immunosuppression and its associated complications.

Brighter Future for Organ Transplantation

The field of organ transplantation has witnessed remarkable advancements in transplant techniques aimed at overcoming rejection and improving transplant outcomes. From targeted immunosuppression to tolerance induction and biomarker monitoring, these innovations offer new hope for transplant recipients and promise to usher in a new era of precision medicine in transplantation.

While challenges remain, including the need for further research, validation, and clinical translation, the future of organ transplantation appears brighter than ever before. With continued investment in research and collaboration between clinicians, scientists, and industry partners, we can overcome rejection and ensure that every transplant recipient has the opportunity for a successful and durable transplant, free from the fear of rejection and its consequences.

Massachusetts General Hospital

Pioneering Transplant History

Massachusetts General Hospital (MGH), founded in 1811, has long been at the forefront of medical innovation and excellence. Throughout its storied history, MGH has played a pioneering role in advancing the field of transplantation, from performing groundbreaking surgeries to developing novel transplant techniques and therapies. This essay explores the rich history of transplantation at Massachusetts General Hospital, highlighting key milestones, notable achievements, and the hospital's ongoing commitment to improving patient care through innovation and collaboration.

1. Early Years and the Birth of Transplantation

In its early years, Massachusetts General Hospital established itself as a leading center for medical education, research, and patient care. The hospital's commitment to excellence attracted some of the brightest minds in medicine, laying the foundation for future advancements in transplantation.

The concept of organ transplantation dates back to the late 19th century, with early experiments in animal-to-animal transplantation laying the groundwork for future clinical applications. MGH played a pivotal role in advancing the field of transplantation, with researchers and clinicians at the hospital contributing to the development of surgical techniques, immunosuppressive therapies, and organ preservation methods.

2. The First Human-to-Human Organ Transplant

One of the most significant milestones in transplantation history occurred at Massachusetts General Hospital in 1954 when Dr. Joseph Murray performed the world's first successful human-to-human kidney transplant. The surgery, which involved transplanting a kidney from one identical twin to another, marked a major breakthrough in the field of transplantation and paved the way for future advancements in organ transplantation.

Dr. Murray's pioneering work at MGH earned him the Nobel Prize in Physiology or Medicine in 1990, cementing the hospital's reputation as a leader in transplantation research and innovation. The success of the first kidney transplant at MGH sparked renewed interest in transplantation worldwide, leading to further advancements in surgical techniques, immunosuppressive therapies, and organ procurement methods.

3. Expansion of Transplant Services

In the decades following the first successful kidney transplant, Massachusetts General Hospital continued to expand its transplant services, offering a wide range of organ transplantations, including liver, heart, lung, pancreas, and bone marrow transplants. The hospital's multidisciplinary team of transplant specialists, including surgeons, physicians, nurses, and allied health professionals, worked collaboratively to provide comprehensive care to transplant recipients and their families.

MGH's commitment to excellence in transplantation was reflected in its outcomes, with the hospital consistently achieving high success rates and excellent patient outcomes. The hospital's transplant program attracted patients from around the world, seeking expert care and innovative treatment options not available elsewhere.

4. Research and Innovation

Research has always been a cornerstone of Massachusetts General Hospital's transplant program, with clinicians and scientists at the hospital conducting groundbreaking research to advance the field of transplantation. From basic science research aimed at understanding the mechanisms of organ rejection to clinical trials evaluating novel immunosuppressive therapies and tolerance induction protocols, MGH has been at the forefront of transplant research and innovation.

The hospital's close collaboration with Harvard Medical School and other academic institutions has facilitated interdisciplinary research and cross-disciplinary collaborations, leading to new discoveries and breakthroughs in transplantation science. MGH's transplant researchers have made significant contributions to our understanding of transplant immunology, organ preservation, and donor-recipient matching, driving improvements in transplant outcomes and patient care.

5. Leadership and Legacy

Throughout its history, Massachusetts General Hospital has been led by visionary leaders who have shaped the direction of the hospital's transplant program and fostered a culture of innovation and excellence. From Dr. Joseph Murray, the father of modern transplantation, to Dr. Joren Madsen, a renowned transplant surgeon and researcher, MGH's transplant leaders have set the standard for excellence in patient care, research, and education.

The legacy of Massachusetts General Hospital in transplantation continues to inspire future generations of transplant clinicians and researchers, who are dedicated to building on the hospital's rich history and advancing the field of transplantation. With its commitment to innovation, collaboration, and compassionate care, MGH remains a global leader in transplantation, providing hope and healing to transplant recipients around the world.

6. Conclusion: A Legacy of Excellence

Massachusetts General Hospital has played a pivotal role in shaping the history of transplantation, from performing the world's first successful human-to-human organ transplant to advancing the field through research and innovation. With its multidisciplinary approach to patient care, commitment to excellence, and legacy of leadership, MGH continues to be a beacon of hope for transplant recipients and a driving force in the quest to improve transplant outcomes and save lives. As the field of transplantation evolves, Massachusetts General Hospital remains steadfast in its mission to push the boundaries of science, deliver exceptional care to patients, and pioneer new treatments and therapies that will transform the future of transplantation.

Genetic Modifications of Pig Donors

Revolutionizing Xenotransplantation for Human Compatibility

Xenotransplantation, the transplantation of organs or tissues from one species to another, holds immense potential to address the critical shortage of donor organs for human transplantation. Among potential donor species, pigs have emerged as a promising candidate due to their physiological similarities to humans and their abundant availability. However, successful xenotransplantation requires overcoming significant immunological barriers, particularly the risk of rejection by the recipient's immune system. Genetic modification of pig donors represents a revolutionary approach to making their organs compatible with human recipients. This essay explores the process of genetic modification in pig donors, its implications for xenotransplantation, and the challenges and opportunities it presents.

1. Understanding Immunological Barriers in Xenotransplantation

Before delving into genetic modifications, it's essential to understand the immunological barriers that must be overcome in xenotransplantation. When a pig organ is transplanted into a human recipient, the recipient's immune system recognizes the pig organ as foreign and mounts a vigorous immune response, leading to rejection. This rejection can occur through various mechanisms, including hyperacute rejection, acute cellular rejection, and chronic rejection, all of which pose significant challenges to the success of xenotransplantation.

2. Genetic Modification Techniques

Genetic modification of pig donors involves the targeted manipulation of the pig genome to reduce the risk of rejection and improve compatibility with human recipients. This process typically involves the use of gene editing technologies, such as CRISPR-Cas9, to introduce specific genetic modifications into pig embryos. One approach to genetic modification involves "knocking out" genes that encode proteins recognized by the human immune system as foreign. For example, pigs can be engineered to lack the expression of alpha-1,3-

galactosyltransferase (α-Gal), an enzyme responsible for the synthesis of a carbohydrate structure known as α-galactose. The presence of α-gal on the surface of pig cells is a major barrier to xenotransplantation, as it triggers an immediate immune response in humans. In addition to knocking out genes, genetic modification can also involve the insertion of human genes into the pig genome to confer immunological tolerance or compatibility. For example, pigs can be engineered to express human complement regulatory proteins or other immune-modulating molecules, which help to dampen the recipient's immune response and promote graft acceptance.

3. Advantages of Genetic Modification

Genetic modification of pig donors offers several advantages for xenotransplantation. By eliminating or modifying genes that trigger immune rejection, genetically modified pigs can produce organs that are less likely to be recognized as foreign by the recipient's immune system. This reduces the risk of rejection and increases the likelihood of graft survival, improving transplant outcomes for patients in need. Furthermore, genetic modification allows for the precise engineering of pig organs to address specific immunological barriers and optimize compatibility with human recipients. This level of customization enables researchers to tailor donor organs to individual patients, potentially reducing the need for immunosuppressive therapy and minimizing the risk of complications associated with transplantation.

4. Challenges and Ethical Considerations

Despite the promise of genetic modification, several challenges and ethical considerations must be addressed before genetically modified pig organs can be used in clinical transplantation. One challenge is the potential for unintended off-target effects of gene editing, which could result in unforeseen consequences for both the donor pig and the recipient. Another consideration is the risk of zoonotic transmission of infectious diseases from pigs to humans. While rigorous screening and monitoring protocols can mitigate this risk, ensuring the safety of xenotransplantation remains a priority for researchers and regulatory agencies. Ethical concerns also surround the use of genetically modified animals for organ donation, including questions about animal welfare, genetic integrity, and the implications of creating animals for human benefit. These concerns underscore the importance of ethical oversight and public engagement in the development and implementation of xenotransplantation therapies.

5. Clinical Applications and Future Directions

Despite these challenges, genetic modification of pig donors holds tremendous promise for the future of xenotransplantation. Clinical trials and preclinical studies have demonstrated the feasibility and safety of using genetically modified pig organs in transplantation, with encouraging results in terms of graft survival and patient outcomes. Looking ahead, researchers are exploring novel approaches to genetic modification, including the use of gene editing technologies to create "universal donor" pigs whose organs can be transplanted into any recipient without the need for immunosuppressive therapy. Other areas of research focus on enhancing the immunological compatibility and functionality of pig organs through the targeted modification of specific genes and pathways.

Genetic modification of pig donors represents a revolutionary approach to overcoming immunological barriers in xenotransplantation. By harnessing the power of gene editing technologies, researchers can create pigs whose organs are more compatible with human recipients, offering new hope for patients in need of life-saving transplants. While challenges and ethical considerations remain, the potential benefits of genetically modified pig organs in transplantation are undeniable, paving the way for a future where organ shortages are a thing of the past, and every patient has access to a compatible donor organ.

Understanding Richard Slayman's Medical Background

Richard "Rick" Slayman's medical background provides a compelling narrative of resilience, perseverance, and the challenges faced by individuals living with chronic illness. From his early diagnosis of type 2 diabetes to his journey through end-stage renal disease and ultimately, his groundbreaking xenotransplantation surgery, Rick's story is a testament to the human spirit and the transformative power of modern medicine. This essay explores Richard Slayman's medical background in depth, shedding light on his experiences, medical history, and the factors that led him to become a pioneer in xenotransplantation.

1. Early Life and Diagnosis of Type 2 Diabetes

Rick Slayman was born and raised in Weymouth, Massachusetts, where he lived a relatively normal life until his early thirties when he was diagnosed with type 2 diabetes. Type 2 diabetes, a chronic metabolic disorder characterized by high blood sugar levels, is often associated with lifestyle factors such as diet, exercise, and genetics. Rick's diagnosis marked the beginning of his journey with chronic illness, requiring him to make significant lifestyle changes and manage his condition with medication and monitoring.

2. Complications and Progression to End-Stage Renal Disease

Despite Rick's efforts to manage his diabetes, the condition progressively worsened over time, leading to complications such as diabetic nephropathy, or kidney damage. Diabetic nephropathy is a common complication of diabetes and a leading cause of end-stage renal disease (ESRD), a condition in which the kidneys fail to function adequately, requiring dialysis or transplantation to sustain life. Rick's progression to ESRD marked a significant turning point in his medical journey, necessitating more intensive treatment and care.

3. Treatment and Challenges of Dialysis

As Rick's kidney function declined, he faced the daunting reality of life on dialysis, a life-sustaining treatment that involves the removal of waste products and excess fluid from the blood. Dialysis, while essential for patients with ESRD, is often burdensome and time-consuming, requiring regular visits to a dialysis center and adherence to strict dietary and fluid restrictions. Rick's experience with dialysis underscored the challenges faced by individuals living with chronic illness and the impact it can have on their quality of life.

4. First Kidney Transplant and Complications

In 2018, Rick received his first kidney transplant from a deceased donor, offering him a new lease on life after years of struggling with kidney failure. The transplant surgery initially appeared successful, and Rick's new kidney functioned well, allowing him to resume many of his daily activities and enjoy an improved quality of life. However, complications arose when Rick's body began to reject the donor kidney, ultimately leading to graft failure and the need for Rick to resume dialysis.

5. Consideration for Xenotransplantation

Faced with the failure of his first kidney transplant and the challenges of dialysis, Rick and his medical team began to explore alternative treatment options, including the possibility of xenotransplantation. Xenotransplantation, the transplantation of organs or tissues from one species to another, offered Rick a potential lifeline and a chance for a new kidney that could bypass the immunological barriers associated with human-to-human transplantation. Rick's consideration for xenotransplantation marked a pivotal moment in his medical journey, requiring careful deliberation and consultation with his medical providers.

6. Decision to Proceed with Xenotransplantation

After careful consideration and consultation with his medical team, Rick made the courageous decision to undergo xenotransplantation with a genetically modified pig kidney. Despite the risks and uncertainties associated with the procedure, Rick saw it as his best chance for survival and a way to pave the path for future patients in need of organ transplants. Rick's decision to proceed with xenotransplantation exemplified his resilience, determination, and unwavering optimism in the face of adversity.

7. Xenotransplantation Surgery and Recovery

On March 16, 2024, Richard Slayman underwent a historic xenotransplantation surgery at Massachusetts General Hospital, becoming the first person to receive a genetically modified pig kidney. The surgery, led by a team of skilled surgeons, was successful, and Rick's new kidney began functioning immediately, producing urine and offering hope for a brighter future. In the days and weeks following the surgery, Rick's condition steadily improved, and he began to show signs of recovery, marking a new chapter in his medical journey and a testament to the power of modern medicine and innovation.

A Journey of Hope and Resilience

Richard Slayman's medical background is a remarkable story of hope, resilience, and the human spirit. From his early diagnosis of type 2 diabetes to his pioneering xenotransplantation surgery, Rick's journey exemplifies the challenges and triumphs faced by individuals living with chronic illness. Through his courage, determination, and unwavering optimism, Rick has become a beacon of hope for patients awaiting life-saving organ transplants and a testament to the transformative power of modern medicine. As Rick continues his recovery and adjusts to life with his new kidney, his journey serves as a reminder of the resilience of the human spirit and the boundless possibilities of medical innovation.

Prognosis After Transplant

Navigating the Journey to Recovery and Beyond

The journey of a transplant recipient does not end with the successful transplantation of a new organ. Instead, it marks the beginning of a new chapter filled with challenges, triumphs, and a renewed sense of hope for the future. Understanding the prognosis after transplant is essential for patients and their caregivers, as it provides insight into what to expect during the recovery process and beyond. This essay explores the prognosis after transplant, including the short-term and long-term outcomes, challenges, and strategies for optimizing health and well-being post-transplant.

1. Short-Term Prognosis: Immediate Recovery and Hospital Stay

Following transplant surgery, patients typically spend several days to weeks in the hospital recovering from the procedure and closely monitored for signs of complications. The immediate post-transplant period is critical for ensuring graft function, managing pain, and preventing infections. Patients may experience side effects from immunosuppressive medications, such as nausea, vomiting, and increased susceptibility to infections. Close communication with the transplant team and adherence to medication regimens are essential during this time to ensure a successful recovery and minimize the risk of complications.

2. Early Complications and Management

While most transplant recipients experience a smooth recovery, some may encounter early complications that require prompt intervention and management. Common complications include surgical complications, such as bleeding, infection, and wound dehiscence, as well as medical complications, such as organ rejection, infection, and adverse reactions to immunosuppressive medications. Early detection and treatment of complications are crucial for preventing graft loss and optimizing long-term outcomes. Close follow-up with the transplant team and adherence to medical recommendations are essential during the early post-transplant period to mitigate the risk of complications and promote healing.

3. Long-Term Prognosis: Graft Survival and Quality of Life

The long-term prognosis after transplant is influenced by various factors, including the type of organ transplanted, the underlying cause of organ failure, the recipient's overall health, and adherence to medical recommendations. While many transplant recipients enjoy long-term graft survival and improved quality of life, some may experience complications or challenges that impact their long-term prognosis.

One of the primary concerns after transplant is the risk of organ rejection, which can occur months to years after transplantation and requires ongoing monitoring and management. Other long-term complications may include infections, cardiovascular disease, metabolic disorders, and malignancies, all of which can affect graft function and overall health.

4. Adherence to Medical Regimens and Lifestyle Modifications

Adherence to medical regimens, including immunosuppressive medications, is essential for maintaining graft function and preventing complications after transplant. Transplant recipients must follow their prescribed medication regimen diligently, attend regular follow-up appointments with their transplant team, and undergo routine laboratory testing to monitor graft function and detect early signs of rejection or complications.

In addition to medication adherence, lifestyle modifications are often recommended to optimize health and well-being after transplant. These may include dietary changes, regular exercise, smoking cessation, and alcohol moderation. By adopting a healthy lifestyle and adhering to medical recommendations, transplant recipients can reduce the risk of complications, improve their overall health, and enhance their long-term prognosis.

5. Psychosocial Support and Mental Health

The psychosocial impact of transplant surgery should not be overlooked, as it can have a significant impact on the recipient's overall well-being and quality of life. Many transplant recipients experience a range of emotions during the recovery process, including anxiety, depression, stress, and feelings of uncertainty about the future. Psychosocial support services, including counseling, support groups, and peer mentoring, can provide valuable emotional support and coping strategies for transplant recipients and their families.

6. Returning to Normal Activities and Resuming Work

Returning to normal activities and resuming work after transplant is an important milestone for many transplant recipients, signaling a return to independence and normalcy. However, the timing and feasibility of returning to work will vary depending on the individual's recovery, the type of transplant, the nature of their job, and any physical or cognitive limitations resulting from the transplant. Transplant recipients should work closely with their transplant team and employer to develop a gradual return-to-work plan that accommodates their medical needs and ensures a safe and successful transition back to the workforce. Occupational therapy and vocational rehabilitation services may be helpful for transplant recipients adjusting to work-related challenges and reintegration into the workforce.

7. Long-Term Follow-Up and Surveillance

Long-term follow-up and surveillance are essential for monitoring graft function, detecting complications, and optimizing long-term outcomes after transplant. Transplant recipients should continue to undergo regular follow-up appointments with their transplant team, including clinical evaluations, laboratory testing, imaging studies, and other diagnostic procedures as needed. Medical surveillance, transplant recipients should be proactive about maintaining their overall health and well-being by adopting healthy lifestyle habits, managing chronic conditions, and seeking timely medical care for any new or worsening symptoms. By staying engaged in their care and adhering to medical recommendations, transplant recipients can maximize their chances of long-term graft survival and enjoy a fulfilling and rewarding life post-transplant.

A Journey of Hope and Healing

The prognosis after transplant is a multifaceted journey characterized by challenges, triumphs, and the transformative power of modern medicine. While the road to recovery may be long and challenging, transplant recipients have the opportunity to reclaim their health, regain their independence, and embrace a new lease on life. By staying informed, engaged, and proactive about their care, transplant recipients can navigate the complexities of post-transplant life with confidence and resilience, knowing that they are supported by a dedicated team of healthcare professionals and empowered to live life to the fullest.

Comparing Success Rates with Previous Attempts

Advancements in Transplantation

Comparing success rates with previous attempts in transplantation provides valuable insights into the evolution of transplant medicine, highlighting advancements, challenges, and the ongoing quest to improve patient outcomes. From the early experiments in organ transplantation to the latest breakthroughs in xenotransplantation, this essay explores the historical context, key milestones, and comparative success rates across different eras of transplant medicine.

1. Early Attempts at Organ Transplantation

The history of organ transplantation dates back to the early 20th century, with pioneering surgeons and researchers laying the groundwork for modern transplant medicine. The first successful organ transplant occurred in 1905 when Dr. Eduard Zirm performed a corneal transplant, restoring vision in a patient with corneal blindness. Subsequent attempts at organ transplantation, including kidney, liver, and heart transplants, faced significant challenges due to technical limitations, immunological barriers, and post-operative complications. Despite early setbacks, these early pioneers paved the way for future advancements in transplantation and laid the foundation for the field's rapid evolution in the decades to come.

2. Development of Immunosuppressive Therapies

One of the key milestones in transplant medicine was the development of immunosuppressive therapies, which revolutionized the field by reducing the risk of rejection and improving graft survival. The discovery of corticosteroids in the 1950s and the introduction of calcineurin inhibitors, such as cyclosporine and tacrolimus, in the 1980s transformed the landscape of organ transplantation, enabling surgeons to perform increasingly complex and successful transplants. These immunosuppressive agents, combined with advances in surgical techniques and organ preservation methods, contributed to significant improvements in transplant outcomes and patient survival rates.

3. Evolution of Organ Procurement and Allocation

Over time, organ procurement and allocation systems have evolved to address the growing demand for donor organs and ensure equitable access to transplantation. The establishment of national organ procurement organizations (OPOs) in the United States in the 1970s standardized organ donation practices and facilitated the efficient recovery and distribution of donor organs. Additionally, the implementation of organ allocation algorithms, such as the Model for End-Stage Liver Disease (MELD) score for liver transplantation, has improved the allocation of organs based on medical urgency and patient need, leading to better outcomes and reduced waiting times for transplant candidates.

4. Advancements in Surgical Techniques

Advancements in surgical techniques have played a crucial role in improving transplant outcomes and expanding the scope of transplantation to include complex multi-organ transplants and innovative procedures. Minimally invasive surgical approaches, such as laparoscopic donor nephrectomy for kidney transplantation, have reduced surgical morbidity and accelerated recovery times for both donors and recipients. Additionally, the advent of robotic-assisted surgery has enabled surgeons to perform intricate transplant procedures with enhanced precision and control, further improving patient outcomes and expanding access to transplantation.

5. Introduction of Xenotransplantation

Xenotransplantation, the transplantation of organs or tissues from one species to another, represents a promising frontier in transplant medicine, offering the potential to address the critical shortage of donor organs and improve transplant outcomes. While early attempts at xenotransplantation faced significant challenges, including hyperacute rejection and immune-mediated graft failure, recent advancements in genetic engineering and immunomodulation have renewed interest in the field and yielded promising results in preclinical and clinical studies. The successful transplantation of genetically modified pig organs into human recipients, as demonstrated by the recent case of Richard Slayman, represents a significant breakthrough in xenotransplantation and underscores the potential of this approach to revolutionize organ transplantation in the future.

6. Comparative Success Rates Across Eras

Comparing success rates across different eras of transplant medicine provides a nuanced perspective on the evolution of transplantation and the impact of advancements in surgical techniques, immunosuppressive therapies, and organ allocation systems on patient outcomes. While early attempts at organ transplantation were met with high rates of graft failure and patient mortality, the introduction of immunosuppressive agents in the mid-20th century significantly improved transplant survival rates and paved the way for the widespread adoption of transplantation as a standard of care for end-stage organ failure. Subsequent refinements in surgical techniques, organ preservation methods, and post-operative care further contributed to improvements in transplant outcomes, with survival rates continuing to improve over time.

7. Challenges and Opportunities for the Future

Despite the remarkable progress made in transplant medicine, significant challenges remain, including the shortage of donor organs, the risk of organ rejection, and the side effects of long-term immunosuppressive therapy. Addressing these challenges will require continued innovation, collaboration, and investment in research to develop novel therapies, such as xenotransplantation and tissue engineering, that offer the potential to overcome the limitations of traditional transplantation. Additionally, efforts to increase organ donation rates, improve organ preservation techniques, and enhance post-transplant care will be essential for further improving transplant outcomes and ensuring equitable access to transplantation

Future Possibilities

Transforming the Landscape for Dialysis Patients

The landscape of renal replacement therapy is on the brink of transformation, offering new hope and possibilities for patients living with end-stage renal disease (ESRD). As advancements in medical science and technology continue to accelerate, the future holds promising opportunities to improve the quality of life, outcomes, and options available to dialysis patients. This essay explores the potential impact of future possibilities on dialysis patients, ranging from innovative therapies to emerging technologies and healthcare delivery models.

1. Next-Generation Dialysis Technologies

The future of dialysis holds the promise of next-generation technologies that aim to enhance the efficiency, convenience, and effectiveness of renal replacement therapy. From wearable and portable dialysis devices to bioartificial kidneys and implantable renal assist devices, researchers and engineers are exploring innovative approaches to dialysis that offer greater flexibility, autonomy, and comfort for patients. These advancements have the potential to revolutionize the dialysis experience, allowing patients to receive treatment on their own terms and integrate dialysis into their daily lives more seamlessly.

2. Bioengineered and Regenerative Therapies

Bioengineering and regenerative medicine hold significant promise for the development of alternative therapies for ESRD that go beyond conventional dialysis. Researchers are exploring the use of stem cells, tissue engineering, and organoid technology to create bioengineered kidneys and other renal tissues that can replace or augment the function of diseased kidneys. Additionally, efforts to develop immunomodulatory and anti-fibrotic therapies aim to prevent or reverse the progression of kidney disease and delay the need for dialysis in high-risk patients. These regenerative approaches offer the potential to restore kidney function, improve quality of life, and reduce the burden of dialysis for patients with ESRD.

3. Personalized Medicine and Precision Nephrology

The future of nephrology lies in personalized medicine and precision nephrology, where treatments are tailored to the individual characteristics, genetics, and underlying pathophysiology of each patient. Advances in genomics, biomarker discovery, and data analytics enable clinicians to identify patients at risk for kidney disease, predict progression, and optimize treatment strategies based on their unique genetic profile and clinical phenotype. By leveraging personalized approaches to care, clinicians can better manage kidney disease, optimize outcomes, and personalize treatment decisions for dialysis patients, leading to improved quality of life and better long-term prognosis.

4. Telehealth and Remote Monitoring

Telehealth and remote monitoring technologies are poised to revolutionize the delivery of care for dialysis patients, offering greater access, convenience, and continuity of care. Virtual care platforms, mobile health apps, and wearable devices enable patients to connect with their healthcare providers remotely, access educational resources, and track their health metrics in real-time. Remote monitoring of vital signs, laboratory values, and treatment adherence allows clinicians to detect early signs of complications, intervene promptly, and optimize therapy to prevent adverse outcomes. These telehealth solutions have the potential to expand access to care, improve patient engagement, and enhance outcomes for dialysis patients, particularly those in underserved or rural communities.

5. Artificial Intelligence and Machine Learning

Artificial intelligence (AI) and machine learning are revolutionizing the field of nephrology by providing clinicians with powerful tools for data analysis, decision support, and predictive modeling. AI algorithms can analyze large datasets, identify patterns, and predict clinical outcomes, helping clinicians make more informed decisions about diagnosis, treatment, and prognosis. Machine learning algorithms can also optimize dialysis prescriptions, individualize fluid management, and predict patient responses to therapy, leading to more personalized and effective care for dialysis patients. By harnessing the power of AI and machine learning, clinicians can improve outcomes, reduce costs, and enhance the overall quality of care for dialysis patients.

6. Patient-Centered Care and Shared Decision-Making

The future of dialysis care is centered around the principles of patient-centered care and shared decision-making, where patients are empowered to actively participate in their care and make informed decisions about their treatment options. Shared decision-making tools, decision aids, and patient decision support resources facilitate meaningful conversations between patients and clinicians, helping patients navigate complex treatment decisions and align their care with their values, preferences, and goals. By involving patients in the decision-making process, clinicians can enhance patient satisfaction, improve treatment adherence, and promote better outcomes for dialysis patients, fostering a collaborative partnership between patients and providers that prioritizes individualized care and holistic well-being.

7. Health Equity and Access to Care

Achieving health equity and improving access to care for dialysis patients is a critical priority for the future of nephrology. Disparities in access to transplantation, dialysis facilities, and preventive care contribute to inequalities in kidney health outcomes, particularly among underserved and marginalized populations. Efforts to address social determinants of health, promote health literacy, and expand access to affordable, culturally competent care are essential for reducing disparities and improving outcomes for all dialysis patients. By advocating for policies that promote health equity and investing in community-based initiatives, stakeholders can work together to ensure that all patients have access to high-quality, comprehensive care that meets their unique needs and

Cost and Approval Considerations in Advanced Medical Therapies

As medical science progresses and novel therapies emerge, the considerations surrounding cost and approval become increasingly complex and significant. This essay delves into the intricate interplay between cost and approval considerations for advanced medical therapies, exploring the challenges, implications, and strategies for navigating this evolving landscape.

1. Understanding the Economics of Advanced Therapies

The economics of advanced medical therapies are multifaceted, encompassing various factors such as research and development costs, manufacturing expenses, clinical trial expenditures, and pricing strategies. Developing innovative therapies often requires substantial investment in research, infrastructure, and regulatory compliance, driving up the overall cost of bringing new treatments to market. Additionally, the complexity and novelty of advanced therapies may pose challenges in terms of manufacturing scalability, supply chain logistics, and quality control, further influencing their cost and affordability.

2. Pricing and Reimbursement Challenges

Determining the appropriate pricing and reimbursement strategy for advanced therapies is a complex and contentious issue that requires careful consideration of multiple stakeholders' interests, including patients, payers, providers, manufacturers, and policymakers. Manufacturers must balance the need for fair return on investment with considerations of affordability, accessibility, and equity, while payers seek to contain costs, manage utilization, and ensure value for money. Striking the right balance between pricing, reimbursement, and patient access is essential for fostering innovation, promoting sustainability, and optimizing patient outcomes.

3. Cost-Effectiveness and Value Assessment

Assessing the cost-effectiveness and value of advanced medical therapies is essential for informing decision-making, resource allocation, and reimbursement policies.

Health technology assessment (HTA) agencies, payers, and policymakers use economic evaluation methods, such as cost-effectiveness analysis, budget impact analysis, and quality-adjusted life years (QALYs), to assess the relative value of different treatments and inform coverage decisions. Incorporating considerations of clinical effectiveness, safety, patient preferences, and societal impact, these evaluations help stakeholders prioritize interventions, allocate resources efficiently, and maximize health gains within finite healthcare budgets.

4. Regulatory Approval Pathways

Navigating the regulatory approval process for advanced medical therapies involves navigating a complex maze of regulations, guidelines, and requirements established by regulatory agencies such as the Food and Drug Administration (FDA) in the United States and the European Medicines Agency (EMA) in Europe. Manufacturers must demonstrate the safety, efficacy, and quality of their products through rigorous preclinical and clinical testing, adherence to good manufacturing practices (GMP), and submission of comprehensive regulatory dossiers. Expedited approval pathways, such as the FDA's Breakthrough Therapy designation and the EMA's Priority Medicines (PRIME) scheme, offer accelerated review and approval for promising therapies addressing unmet medical needs, facilitating faster access to patients in need.

5. Health Technology Assessment and Reimbursement

Health technology assessment (HTA) plays a crucial role in informing reimbursement decisions for advanced medical therapies, providing evidence-based evaluations of their clinical effectiveness, cost-effectiveness, and value for money. HTA agencies evaluate the comparative clinical and economic benefits of new therapies relative to existing standards of care, considering factors such as efficacy, safety, patient-reported outcomes, and budget impact. These assessments inform reimbursement policies, coverage decisions, and pricing negotiations between manufacturers and payers, helping to ensure that patients have access to high-value treatments that offer the greatest clinical benefit at an acceptable cost.

6. Challenges in Assessing Value

Assessing the value of advanced medical therapies presents several challenges, including methodological complexities, uncertainty surrounding long-term outcomes, and ethical considerations. Traditional HTA frameworks may not fully capture the unique characteristics and benefits of advanced therapies, such as their potential to cure or significantly improve outcomes for patients with debilitating or life-threatening conditions. Additionally, measuring value in terms of QALYs or other standard metrics may not adequately reflect the full spectrum of patient preferences, values, and experiences, raising questions about the appropriateness of cost-effectiveness thresholds and reimbursement criteria for advanced therapies.

7. Addressing Affordability and Access

Ensuring affordability and access to advanced medical therapies is a critical priority for policymakers, healthcare providers, and patient advocacy groups, particularly in light of rising healthcare costs, growing budget constraints, and widening health disparities. Strategies to enhance affordability and access may include price negotiations, value-based pricing agreements, risk-sharing arrangements, patient assistance programs, and alternative financing models such as subscription-based payment models and outcomes-based reimbursement. By implementing innovative financing and reimbursement strategies, stakeholders can enhance patient access to advanced therapies while promoting sustainability, equity, and value in healthcare delivery.

8. The Role of Real-World Evidence

Real-world evidence (RWE) plays an increasingly important role in complementing traditional clinical trial data and informing decision-making about the value, effectiveness, and safety of advanced medical therapies in routine clinical practice. RWE derived from electronic health records, registries, claims data, and patient-reported outcomes provides valuable insights into treatment patterns, healthcare utilization, and patient outcomes in real-world settings, helping to bridge the gap between clinical research and clinical practice. By incorporating RWE into HTA and reimbursement processes, stakeholders can make more informed decisions about the adoption, utilization, and coverage of advanced therapies, ultimately improving patient care and health outcomes.

The intersection of cost and approval considerations in advanced medical therapies represents a complex and dynamic landscape shaped by scientific innovation, regulatory requirements, economic constraints, and patient needs. Navigating these challenges requires collaboration, evidence-based decision-making, and a commitment to ensuring that patients have timely access to safe, effective, and affordable treatments that improve health outcomes and enhance quality of life. By addressing the interrelated issues of pricing, reimbursement, value assessment, and access, stakeholders can foster a healthcare system that promotes innovation, sustainability, and equity, while meeting the evolving needs of patients in an era of rapid medical advancement.

Looking Ahead in Transplant Medicine

Embracing the Future of Transplant Medicine

As we embark on the final leg of our journey through the landscape of transplant medicine, we find ourselves at a pivotal moment in history, poised on the cusp of unprecedented advancements and transformative breakthroughs. In this concluding chapter, we cast our gaze towards the horizon, envisioning a future where the promise of transplantation shines brighter than ever before. From regenerative therapies to precision medicine, from xenotransplantation to artificial organs, the possibilities are endless as we dare to imagine a world where every patient in need has access to life-saving treatments and where the burden of organ failure is alleviated through innovation, compassion, and collaboration.

1. Harnessing the Power of Regenerative Medicine

Regenerative medicine holds immense potential for revolutionizing the field of transplantation, offering a paradigm shift from organ replacement to organ regeneration. Stem cell therapy, tissue engineering, and organoid technology are paving the way for the development of bioengineered organs and tissues that can be custom-tailored to the individual needs of patients. By harnessing the regenerative capacity of cells and tissues, researchers are unlocking new possibilities for repairing, replacing, or regenerating damaged organs, reducing the reliance on donor organs and expanding access to transplantation for patients in need.

2. Advancing Xenotransplantation as a Viable Option

Xenotransplantation, the transplantation of organs or tissues from one species to another, represents a promising frontier in transplant medicine, offering a potential solution to the critical shortage of donor organs. Recent advancements in genetic engineering, immunomodulation, and organ preservation have paved the way for successful xenotransplantation in preclinical and clinical studies, demonstrating the feasibility and safety of transplanting organs from genetically modified animals into humans. As researchers continue to refine techniques and address remaining challenges, xenotransplantation holds the potential to revolutionize organ transplantation and improve outcomes for patients awaiting life-saving transplants.

3. Embracing Precision Medicine and Personalized Therapies

The era of precision medicine and personalized therapies is upon us, offering tailored treatment approaches that account for the individual characteristics, genetics, and immunological profiles of transplant recipients. Advances in genomics, biomarker discovery, and immune profiling enable clinicians to identify patients at risk for rejection, predict treatment responses, and optimize therapy based on their unique genetic makeup. By harnessing the power of personalized medicine, clinicians can maximize the efficacy of transplant treatments, minimize the risk of complications, and improve long-term outcomes for patients, ushering in a new era of precision transplantation.

4. Fostering Collaboration and Innovation

Looking ahead, collaboration and innovation will be essential for driving progress and overcoming the remaining challenges in transplant medicine. Multidisciplinary teams of researchers, clinicians, engineers, and industry partners must come together to tackle complex problems, share knowledge, and translate scientific discoveries into clinical applications. By fostering a culture of collaboration and innovation, we can accelerate the pace of progress, break down barriers to innovation, and bring life-saving treatments to patients in need.

5. Addressing Ethical and Societal Considerations

As we chart the course for the future of transplant medicine, we must also grapple with ethical and societal considerations that arise from advancements in the field. Questions surrounding organ procurement, allocation, and consent, as well as the implications of genetic engineering and xenotransplantation, require thoughtful deliberation and engagement with stakeholders. By addressing these ethical and societal considerations head-on, we can ensure that the benefits of transplant medicine are equitably distributed, that patients' rights and dignity are respected, and that we uphold the highest standards of ethical conduct in our pursuit of medical progress.

The future of transplant medicine holds boundless promise, offering hope and healing to millions of patients around the world. From regenerative therapies to precision medicine, from xenotransplantation to artificial organs, the possibilities are limitless as we embark on this journey of discovery and innovation. By harnessing the power of science, embracing collaboration, and upholding our commitment to ethical and societal values, we can create a future where every patient in need has access to life-saving treatments and where the promise of transplantation becomes a reality for all. Together, let us embrace the future of transplant medicine and usher in a new era of hope, healing, and possibility.

Acknowledgment

We extend our heartfelt gratitude to all those who have contributed to the advancement of transplant medicine, shaping its past, present, and future.

First and foremost, we express our deepest appreciation to the patients and their families who have entrusted us with their care and participated in research, clinical trials, and advocacy efforts. Your resilience, courage, and generosity inspire us every day and drive our relentless pursuit of better treatments and outcomes.

We are immensely grateful to the dedicated healthcare professionals, including surgeons, physicians, nurses, and allied health professionals, who work tirelessly to provide compassionate, expert care to transplant patients around the world. Your expertise, compassion, and commitment to excellence are the cornerstone of transplant medicine, and we honor your unwavering dedication to improving patient lives.

We also acknowledge the invaluable contributions of researchers, scientists, and innovators who push the boundaries of knowledge and innovation in transplant medicine. Your groundbreaking discoveries, pioneering technologies, and transformative therapies have revolutionized the field and opened new avenues for improving patient care and outcomes.

Furthermore, we thank the transplant organizations, advocacy groups, and philanthropic supporters who tirelessly advocate for patient rights, raise awareness, and advance research in transplant medicine. Your advocacy, generosity, and collaboration are instrumental in driving progress and addressing the unmet needs of transplant patients worldwide.

Last but not least, we recognize the support of policymakers, regulatory agencies, and healthcare institutions that create an enabling environment for innovation, research, and patient care in transplant medicine. Your leadership, vision, and commitment to advancing healthcare excellence are essential for shaping a future where every patient has access to life-saving transplant therapies. Together, we stand united in our mission to transform the landscape of transplant medicine, improve patient outcomes, and create a brighter future for all those affected by organ failure.

www.ingramcontent.com/pod-product-compliance
Lightning Source LLC
Chambersburg PA
CBHW050310230526
45471CB00005B/2109